Echo Spot Manual
Echo Spot User Guide
By Emery H. Maxwell

Disclaimer:
The views expressed within this book are those of the author alone. *Echo Spot* and *ALEXA* are trademarks of *Amazon*. All other copyrights and trademarks are properties of their respective owners. The information contained within this book is based on the opinions, observations, and experiences of the author and is provided "AS-IS". No warranties of any kind are made. Neither the author nor publisher are engaged in rendering professional services of any kind. Neither the author nor publisher will assume responsibility or liability for any loss or damage related directly or indirectly to the information contained within this book.

The author has attempted to be as accurate as possible with the information contained within this book. Neither the author nor publisher will assume responsibility or liability for any errors, omissions, inconsistencies, or inaccuracies.

Table of Contents

Welcome

Welcome to the *Echo Spot Manual*. This user guide is intended to help you understand, set up, and manage *Echo Spot*.

Since the *Echo Spot* device has a screen, you will be able to see music lyrics, photos, security cameras, watch video flash briefings, and more.

This guide will cover:

• How to set up *Echo Spot*

• How to navigate *Echo Spot*

• How to customize *ALEXA*

• How to teach *ALEXA* your voice

• How to connect to *BLUETOOTH*

• How to enable and manage *Skills*

• How to use calling and messaging

• How to connect and manage *smart home* devices

• How to use *Echo* with *IFTTT*

• How to enable *Voice-View Screen Reader*

• How to enable and use *Screen Magnifier*

• Troubleshooting

• . . . and more.

It's time to get started.

Getting Started

To help you understand the *Echo Spot* device, this section will cover the specifications of *Echo Spot*, basic hardware, visual indicators and their meanings, and how to set up the device.

Specifications

Inside the Box: *Echo Spot* device, cable/power adapter, quick start guide.

Size: 4.1" x 3.8" x 3.6"

Weight: 14.8 oz.

Screen Size: 2.5"

Speaker Size: 1.4"

ALEXA App: The *ALEXA* App is compatible with *Android, IOS,* and *Fire OS* devices. It is also accessible through your browser.

Audio: Built-in speaker. 3.5 mm stereo audio output.

WI-FI Connectivity: Dual-band WI-FI supports 802.11 a/b/g/n networks. Does not support connecting to peer-to-peer WI-FI networks.

BLUETOOTH Connectivity: Advanced Audio Distribution Profile (A2DP) support for audio streaming from your mobile device to *Echo Show* or from *Echo Show* to your *BLUETOOTH* speaker. Audio/Video Remote Control Profile for voice control of connected mobile devices. Hands-free voice control is not supported for *Mac OS X* devices. *BLUETOOTH* speakers requiring PIN codes are not supported.

External Display (Basic Hardware)

Light Ring: Appears on the *Echo Spot* screen and has different meanings reflected by color.

Mic/Camera on/off button and LED: Disables the microphones and camera. It is located at the top of the device on the left side. The LED on the front of the device will turn red when the microphones and camera are turned off. Pushing this button will power on the device. Holding this button down will prompt you to turn off the device.

Touch Screen: The 2.5" screen provides visual information and allows you to interact with the device by touching the screen.

Volume buttons: The plus button increases the volume and the minus button decreases the volume.

Power Port: Located at the back of the device, near the bottom.

3.5 mm Audio Output: Located at the back of the device, next to the power port.

Microphones: 4 dots surrounding the top center of the device.

Front-facing camera: Located on the front side of the device, near the top.

Visual Indicators and Their Meanings

Red: Device's microphones and camera are powered off using the Mic/Camera on/off button on the device.

Blue edges with greenish center: *ALEXA* is listening to your request.

Blue: *ALEXA* is processing your request.

Orange: The device is experiencing connectivity issues.

Purple: *Do Not Disturb* is on.

Setting Up *Echo Spot*

Start by making sure the device is at least eight inches away from any walls and windows.

It should also be kept away from microwaves and other similar devices that might cause interference.

1.) Download the *ALEXA* app.

To download the *ALEXA* app, your mobile device must be one of the following:

• *Fire OS* 3.0 or higher

• *IOS* 9.0 or higher

• *Android* 5.0 or higher

The *ALEXA* app is available at the app store on your mobile device. Simply go to the app store on your mobile device and search for *"ALEXA* app.*"*

If you'd like to use a computer browser to begin the setup process, there is also the option to go to the *ALEXA/Amazon* website to download the app from there.

The browser must be one of the following:

• Internet Explorer (10 or higher)

• Safari

• Chrome

• Microsoft Edge

• Firefox

After the app is downloaded, follow the prompts and sign in.

2.) Power on the *Echo Spot* device.

Plug the included cable into the device and connect it to a power outlet. Then press the *On* button on the *Echo Spot* device.

3.) Follow the Prompts.

There will be a series of prompts that will bring you through the setup process, including language selection and WI-FI connection.

If you are having connection issues, restart the router by unplugging it and plugging it back in.

When selecting your WI-FI network, you might be prompted to enter a password. The password can be found on your router.

If you still can't find the password, contact your Internet Service Provider.

If your network doesn't show up in the list, go to **Add a Network** and type in the information. Then select **Done**.

4.) Read the terms and conditions, then watch the brief *Echo Spot* intro video.

5.) Speak to *ALEXA*.

Screen Settings

Accessing the *Echo* screen settings can be done by swiping down from the upper portion of the screen and selecting **Settings**, or you can say, "Go to settings."

Here are the settings and their descriptions:

BLUETOOTH: Connect your device to an external speaker or *BLUETOOTH* headphones.

WI-FI: View and connect to available networks.

Home Screen / Home & Clock: Select which pages you'd like to see on your home screen, photos backgrounds, themes, and auto-rotation preferences.

Display: Change the level of screen brightness and toggle adaptive brightness and the screen clock.

Sounds: Adjust the volume for different features.

Do Not Disturb: Enabling this will block alerts for calls and messages.

Device Options: Allows you to modify the name of device, location, time, and remote pairing.

Restrict Access: Allows you to block or modify access to certain features on the device.

Things to Try: View examples of how to interact with *ALEXA*.

Help: Find guidance for the device.

Accessibility: Make your device easier to use.

Legal & Compliance: Review content related to legal and compliance.

Once *ALEXA* is set up, you can say, "Turn off the screen," to power off the screen. To power the screen back on, touch the screen or use the device's wake word.

Navigating *Echo Spot*

The *Echo Spot* device can be navigated by voice and the touch screen.

Go to the home screen

- Say, "Go home."
- Using a downward motion, swipe the top of the *Echo Spot* screen, then select **Home**.

View available settings

- Say, "Go to settings."
- Using a downward motion, swipe the upper portion of the *Echo Spot* screen and select **Settings**.

Scroll through a list

- Say, "Scroll up."
- Say, "Scroll down."
- Swipe the screen accordingly.

Media playback

- Say, "Next."
- Say, "Pause."
- Say, "Resume."
- Say, "Previous."
- Say, "Go back."
- Say, "Rewind / fast forward [number] [minutes / seconds / hours]."

Video Navigation

Search video library for movies and TV shows

- "Show me my video library."

- "Show me my watch list."

Find a specific title

- "Show me [title]."

- "Search for [TV series name]."

Search for genre or actor

- "Show me [genre]."

- "Show me [actor's name] movies."

Playback

- "Watch [movie / show title]."
- "Play the movie [title]."
- "Watch [TV series name], season [number]."

Control

- "Next video."
- "Next episode."
- "Fast forward."
- "Rewind."
- "Pause."
- "Resume."
- "Go back [number] [seconds / minutes / hours]."
- "Skip ahead [number] [seconds / minutes / hours]."

Photos Navigation

P rime members receive free unlimited photo storage.

- "Show my photos."
- "Show Family Vault photos."
- "Show this day photos."
- "Show my photo albums."
- "Show my [album title] album."
- "Pause slideshow."
- "Resume slideshow."
- "Next photo."
- "Previous photo."
- "Turn shuffle on."
- "Repeat."
- "Show photos of [name]."
- "Show photos of [place]."

Setting a Background on *Echo Spot*

P hotos can be set as a background on *Echo Spot*.

Set a single photo as a background

1.) Find a photo you'd like to use as a background.

2.) Say, "Set this photo as my background."

Or go to **Settings** in the *ALEXA* app and select your device. From there, go to **Home Screen Background** and select **Choose a photo**. Follow the on-screen prompts.

Set a collection of photos as a background

1.) Find an album you'd like to use as a background.

2.) Say, "Set this album as my background."

Or go to **Settings > Home Screen > Background > Prime Photos > Change**. Then select the album you'd like to set as a background.

How to Customize *ALEXA*

C ustomizing *ALEXA* to match your preferences will allow you to:

• Change your device's location

• Change the Wake word

• Get sports updates

• Shop

• Check the weather forecast

• . . .and more.

Change Device Location

There is are two options to change the device's location: Changing it from the *ALEXA* app or changing it on the device directly.

Change Device Location in the *ALEXA* app

1.) Go to the menu and select **Settings**.

2.) Select your device.

3.) Go to **Device Location**.

4.) Select **Edit**.

5.) Enter your address.

6.) Select **Save**.

Change Location on the Device Directly

1.) Using a downward motion, swipe the top of the screen and select **Settings**. Or say, "Go to settings."

2.) Select **Device Options**.

3.) Select **Device Location**.

4.) Using the keyboard, enter or change the address.

Once your device's location is set, you can now ask about the weather.

Here are some things you can say:

• "What's the weather?"

- "What's the weather for this week?"
- "Will it rain/snow tomorrow?"
- "What's the weather like for [day]?"

Checking Traffic

Checking traffic involves setting and saving your commute information in the *ALEXA* app, then asking, "How is traffic?" You can also say, "What's traffic like right now?" or "What's my commute?"

1.) In the *ALEXA* app, go to menu and select **Settings**.

2.) Go to the **Accounts** section.

3.) Select **Traffic**.

4.) Enter the starting and destination points.

5.) Select **Save Changes**.

Shopping

When placing orders, *ALEXA* uses the default shipping address and payment settings as listed in your account. Aside from having an account and payment method in place, you will also need to be a *Prime* member on *Amazon* to shop with *ALEXA*.

Digital music that is purchased will be stored in your library and will be available for playback on any device that supports *Amazon Music*.

To shop for a song or album, say, "Shop for the song [title]," or "Shop for the album [title]."

To purchase or add a currently playing sample, say, "Buy this [song/album]," or "Add this [song/album]."

Other products can also be ordered via *ALEXA* as long as they are *Prime-eligible*.

Voice Purchasing will be enabled by default once your *ALEXA* device is registered. To manage your voice purchasing settings in the *ALEXA* app, simply go to **Settings**, scroll down to **Accounts**, and then select **Voice Purchasing**.

Changing the Wake Word

The wake word can be changed via the *ALEXA* app or on the device directly.

The available wake words are:

• *ALEXA*

• Computer

• *Echo*

• *Amazon*

Change Wake Word in the *ALEXA* App

1.) Go to the menu and select **Settings**.

2.) Select your device.

3.) Select **Wake Word**.

4.) Using the drop-down menu, select a wake word.

5.) Select **Save**.

Once the wake word is changed, the light ring on your device will briefly turn orange.

Change Wake Word on the Device Directly

1.) Using a downward motion, swipe the upper portion of the screen. Or say, "Go to Settings."

2.) Select **Device Options**.

3.) Select **Wake Word**.

Sports Updates

It's possible to ask *ALEXA* about the latest scores and upcoming games for supported sports teams by saying, "What's my sports update?"

Supported Leagues

- NHL
- MLS
- MLB
- NBA
- NFL
- EPL
- FA Cup
- NCAA Men's Basketball
- NCAA FBS football
- UEFA
- WNBA
- German BUNDESLIGA 1st Division
- German BUNDESLIGA 2nd Division

Add sports teams in the *ALEXA* app

1.) Go to the menu and select **Settings**.

2.) Go to **Accounts**.

3.) Select **Sports Update**.

4.) Using the search bar, search for sports teams.

5.) Select a team to add.

Tip: If you'd like to remove a team, select the **X** next to the team name.

How to Teach *ALEXA* Your Voice

T eaching *ALEXA* to recognize your voice can help create a more personalized experience across certain supported features.

This can be achieved by creating a voice profile.

Create a Voice Profile

1.) On your *smart phone*, go to the menu and select *Settings*.

2.) Go to the *Accounts* section.

3.) Select **Your Voice**.

4.) Select **Begin**.

5.) Using the drop-down menu, select the device you want to interact with to teach *ALEXA* your voice.

6.) Select **Next**.

7.) When prompted, say the phrase out loud. Then select **Next** to go to the next phrase. You can also try the phrase again by selecting **Try Again**.

8.) Select **Complete**.

You should now see a confirmation page on the screen.

To confirm, ask *ALEXA*, "Who am I?" If the process went well, *ALEXA* will mention your name.

<u>How to Connect Your Device to BLUETOOTH</u>

B efore you begin, make sure you are using certified speakers that support *BLUETOOTH* profiles for *Echo* devices.

Supported BLUETOOTH Profiles:

• Advanced Audio Distribution Profile

• Audio / Video Control Profile

When you're ready to get started, follow the steps below.

1.) Power on the BLUETOOTH speaker and make sure the volume is not turned down too far.

2.) Since *Echo* can only connect to one BLUETOOTH device at a time, make sure other BLUETOOTH devices are disconnected from the *Echo* device.

3.) On the *BLUETOOTH* speaker, make sure *Pairing Mode* is on. If you are not sure how to enable or disable *Pairing mode*, check the *BLUETOOTH* speaker's owner's manual.

4.) In the *ALEXA* app, select *Settings*.

5.) Select your device, then select **BLUETOOTH**. Select **Pair a New Device.**

You're *Echo* device should now enter pairing mode.

The speaker will appear in the list of available devices in the *ALEXA* app when *Echo* finds and recognizes your *BLUETOOTH* speaker.

6.) Select your *BLUETOOTH* speaker.

Your *Echo* device will now connect to the speaker.

7.) To finish pairing your *Echo* device with your BLUETOOTH speaker, select *Continue* in the *ALEXA* app.

Once the device is paired, tell *ALEXA* to "Connect." The *Echo* device will connect to the device it was most recently paired with.

How to Enable and Manage *Skills*

S kills are voice-controlled capabilities that improve the *ALEXA* device's functionality. To illustrate, if you'd like *ALEXA* to tell you about specific upcoming events in your city, you would need to enable a specific skill for that.

Oftentimes, if you know the specific name of the skill you'd like to use, you can simply say, "*ALEXA*, enable [skill name]."

But sometimes certain skills need to be enabled through the *Amazon* website or the *ALEXA* app, while others might need to be activated by following the prompts from *ALEXA*.

Enable *Skills*

1.) Open the *ALEXA* app.

2.) Go to the menu and select **Skills**.

You can also go to the *Amazon* website and go into the *skills* section.

3.) Use the *search bar* to find a specific skill or browse through the skills by category.

4.) After finding the skill you'd like to use, select it to go to its detail page. The detail page should include at least one example of what to say to play or open the skill.

5.) On the skill's detail page, select **Enable Skill**.

Now you should be able to tell *ALEXA* to open the skill.

If you need help with the skill, say, "ALEXA, [skill name] help."

Manage *Skills*

1.) Open the *ALEXA* app.

2.) Go to the menu and select **Skills**.

3.) Select **Your Skills**.

4.) Select a *skill* to go to its detail page.

You should now see a list of available options.

ALEXA Calling and Messaging

With *ALEXA Calling and Messaging*, you can make and receive calls and messages between compatible *Echo* devices or the *ALEXA* app.

Most mobile and landline numbers in North America can be called, and it is available to anyone who has access to your compatible devices.

Anyone who has your information and chooses to use the feature can contact you on your compatible *Echo* device or *ALEXA* app.

It is a free service, but signing up is required.

Compatible Devices

• *ALEXA* app

• *Amazon Echo* (1st and 2nd Generation)

• *Echo Dot* (1st and 2nd Generation)

• *Echo Show*

• *Echo Spot*

• *Echo Plus*

How to Sign Up

In order to sign up for *ALEXA Calling and Messaging*, you will need:

• An *Amazon* account

• The *ALEXA* app installed on a compatible *Android* (OS 5.0 or higher) or *IOS* phone (9.0 or higher)

Currently, *ALEXA Calling and Messaging* is only available on *smart* phones. Tablets will not work with this service.

Sign up and set up the calling and messaging feature

1.) Open the *ALEXA* app on your compatible *IOS* or *Android* phone.

2.) Open the *Conversations* tab at the bottom of the menu.

3.) Following the instructions, enter you mobile phone information.

If done correctly, it will import your list of contacts for you.

You might be prompted to verify your mobile number through a text message.

Although *ALEXA Calling and Messaging* is considered separate from your phone service, the *ALEXA* app might use data if your phone is connected to the internet through your mobile network.

How to Use *Echo* to Make Calls

To make a call using the *Echo* device, simply ask *ALEXA* to call the person you want to reach, mentioning the contact's name.

To make a call to the *ALEXA* app or another *Echo* device, say:

"ALEXA, call [person's name] Echo."

To make a call to a landline or mobile number that is saved to your list of contacts, say:

"ALEXA, call [person's name] on his/her home phone."

"ALEXA, call [person's name] mobile."

"ALEXA, call [person's name] at work."

"ALEXA, call [person's name] office."

To verbally dial a mobile or landline without saying the person's name: (Available on *Echo* devices only)

"ALEXA, call [say each digit, including the area code]."

To control the volume, say:

"ALEXA, turn the volume up / down."

You can also mute the line manually by using the **Microphone off** button on the device.

To hang up, say:

"Hang up."

"End call."

If you are making the call from the *ALEXA* app, you can also select the on-screen **End** tab to disconnect.

At this time, *ALEXA* does not support calls to certain types of numbers, such as:

• Emergency services

• Premium-rate or toll numbers

• International numbers outside of North America

• Dial-by-letter numbers

• Abbreviated dial codes

How to Make Calls in The *ALEXA* app

There is also the option to call your *ALEXA* – to – *ALEXA* contacts from the *ALEXA* app.

1.) Open the **Conversations** tab at the bottom of the menu.

2.) Select the **Contacts** icon,

3.) You should now be able to view your list of contacts. Select the contact you would like to call.

4.) Select the *phone* icon to place the call.

Since *Echo Spot* has a screen, the *video* icon can also be used. This will allow the user to place a video call.

To end a call from the *ALEXA* app, select the **End** button.

How to Answer or Ignore Calls

When a call comes in, a green light will flash on the supported *Echo* device. *ALEXA* will also alert you, announcing the caller's contact name.

ALEXA might ask you if you'd like to answer the call. But you can also say, "Answer," or "Ignore."

The green lighting will remain on the device for as long as the call is connected.

If another call comes in while you are still on the line with a different caller, the new call is automatically sent to another device.

If you'd like to temporarily block alerts for calls and messages on the *Echo* device, use the **Do Not Disturb** feature, and say, "Do not disturb me." To switch off the feature, say, "Turn off Do Not Disturb."

The **Do Not Disturb** feature can also be scheduled.

To schedule **Do Not Disturb** in the *ALEXA* app:

1.) Select *Settings* from the menu.

2.) Select your device.

3.) Look under **Do Not Disturb**, then select **Scheduled**.

4.) Using the toggle switch, enable the feature.

5.) Select **Edit**.

6.) You should now be able to edit the start and stop times. Select **Save Changes** when you're done.

How to Use *ALEXA* Messaging

Although *ALEXA* can't send photos or other attachments, it can deliver messages to the recipient's *ALEXA* app and other supported *Echo* devices.

Send Messages in The ALEXA app

1.) Select the *Conversations* icon.

2.) Select the *New Conversations* tab.

3.) You should now be able to see your list of contacts. If you'd like to start a new message, select a contact from the list. You can also continue an existing conversation by selecting the conversation that appears.

4.) Select the text tab to open the keyboard.

5.) When you're done typing the message, select the *Send* tab.

Check Messages in The ALEXA app

1.) Tap the *Conversations* icon.

2.) Select the new message with the green dot.

How to Send and Listen to Voice Messages with The *ALEXA* App

V oice messages can be sent and received through the ALEXA app.

The recipient's *Echo* device will make a chiming sound and turn yellow when a message is received.

Send Voice Messages From ALEXA App

1.) Select the *Conversations* icon at the bottom of the menu.

2.) Select the *New Conversations* tab and select a contact from the list. You can also respond to an existing conversation by selecting from the list of conversations shown.

3.) Press and hold the *Microphone* tab.

4.) Continuing to hold the *Microphone* tab, record your message.

5.) When you're done recording the message, release the *Microphone* tab. The message should now be sent.

Listen to Voice Message in The ALEXA app

1.) Select the *Conversations* icon at the bottom of the menu.

2.) Select the message.

3.) Press the *Play* icon. You can also read the message if it's transcribed into text.

How to Send and Listen to Voice Messages with The *Echo* Device

V oice messages can also be sent and listened to through a supported *Echo* device.

Send Voice Message Using *Echo*

1.) Say, "Send a message to [name of contact]."

2.) For confirmation, *ALEXA* might repeat the name back to you.

3.) Confirm the contact's name.

4.) *ALEXA* should now prompt you for the message.

5.) *ALEXA* will send your message when you're done talking.

Play Voice Messages on *Echo* Device

The recipient's *Echo* device will make a chiming sound and turn yellow when a message is received.

To have the messages played out loud, say, "Play my messages."

If there aren't any new messages, *ALEXA* will ask you if you'd like to hear the old ones.

For multiple household members, say, "Play messages for [household member's name]."

How to Manage Your Contacts and Settings

Managing your *ALEXA Calling and Messaging* settings and *ALEXA – to – ALEXA* contacts can be done from the *ALEXA* app.

How to Manage Your Profile

1.) Tap the *Conversations* tab at the bottom of the menu.

2.) Tap the *Contacts* icon.

3.) Select *My Profile.*

There should now be a list of options, including **Name, Drop In,** and **Caller ID**.

• **Name:** Allows you to change your profile name.

• **Drop In:** Enables you to allow or refuse permission to Drop In for a specific contact. Use the toggle switch to enable or disable this feature.

• **Caller ID:** Allows you to select whether or not you want ALEXA to display your contact information when making calls. Use the toggle switch to enable or disable this feature.

How to Manage Your Contacts

Although the *ALEXA* app does not display all of the contacts saved to your phone, it does display your *ALEXA - to - ALEXA* contacts.

Depending on the type of phone you have, go to *Contacts* or *Address Book.* From there, you should be able to add a contact or edit existing ones.

How to Block an *ALEXA - to - ALEXA* Contact

1.) Open the *ALEXA* app.

2.) Tap the *Conversations* icon at the bottom of the menu.

3.) Scroll down and tap **Block Contacts**.

4.) Select the contact you would like to block.

5.) Select **Block**, then confirm.

How to Set Up Profiles For Multiple Household Members

1.) Tap the *Conversations* icon at the bottom of the menu.

2.) Select **Get Started**.

3.) Add and verify your mobile number.

4.) When you reach the **Help ALEXA get to know you** page, you will have the option to select a profile you have already saved. If you do not have a profile saved, select **I'm someone else**.

5.) Enter your name and mobile phone information.

Each additional user can follow these same steps.

<u>How to Connect *Smart* Home Devices to *ALEXA*</u>

Before connecting a *smart home* device to *ALEXA*, it's important to be familiar with the safety information.

<u>Safety Guidelines</u>

• Follow the instructions for *smart home* devices.

• After a request is made, confirm the action has been completed on the *smart home* device.

• Make sure your *ALEXA* supported device and connected products are running efficiently. For example, make sure the *lock doors* feature is working before you leave your home.

<u>How to Connect a *Smart Home* Device to *ALEXA* in The *ALEXA* App</u>

1.) Go to the menu and select **Skills**.

2.) Search for and locate the skill you are looking for, then select **Enable**.

3.) Follow the on-screen directions to get through the linking process.

4.) Tell *ALEXA* to discover the device by saying "Discover my devices." You can also go into the **Smart Home** section in the *ALEXA* app and select **Add Device**.

<u>How to Discover *Smart Home* Devices without a skill</u>

Not all devices require a skill to connect to *ALEXA*.

If you'd like to connect these devices, simply tell *ALEXA* to "Discover devices."

Some devices might need to be powered on before they can be discovered. For example, if you are using a *Phillips Hue Bridge*, press the button on the bridge before attempting to discover the device.

<u>How to Manage Connected *Smart Home* Devices</u>

It is possible to edit a device's name, disable a device, or delete a device.

1.) Go to the menu in the *ALEXA* app.

2.) Tap **Smart Home**.

3.) Select **Devices**.

4.) Select you *smart home* device.

5.) Select **Edit**.

If you'd like to disable all devices associated to a specific skill, rather than delete the devices one at a time, you can simply disable the skill.

Say, "Disable [skill's name] skill."

You can also:

1.) Go to the menu in the *ALEXA* app

2.) Select **Skills**

3.) Select **Your Skills**

4.) Select a skill.

5.) You should now see a *skill* detail page. Select the *Disable* tab.

<u>Shades and Colors for *Smart* Light Control</u>

It is possible to use *ALEXA* to change the colors and shades of white for light bulbs, as long as the light bulbs are compatible.

<u>To see if your *smart home* light bulb supports colors, use the *ALEXA* app.</u>

1.) Go to the menu and select **Smart Home**.

2.) Select the compatible light bulb.

3.) Select **Edit**.

4.) Check the **About** section. You should then be able to see all the supported color capabilities.

When it's good to go, simply say, "*ALEXA,* make the [*smart home* device / group name] warmer / cooler to incrementally adjust shades of white.

Or say, "*ALEXA,* set the lights to [color]."

<u>*Smart Light colors can also be changed through the ALEXA app*</u>

1.) Go to the menu and select **Smart Home**.

2.) Select a compatible light bulb.

3.) Select **Set Color**.

<u>Shades of White</u>

• White

• Soft White

• Daylight

• Cool White

• Warm White

<u>Colors</u>

• Gold

• Crimson

• Cyan

• Blue

• Green

• Lavender

• Red

- Salmon
- Yellow
- Violet
- Sky Blue
- Teal
- Turquoise
- Pink
- Orange
- Lime
- Purple

There are many more color options that can be viewed via the *ALEXA* app when you go to **Set Color**.

How to Use *Echo* with *IFTTT*

*I*FTTT stands for *If This Then That*. It's an online service that uses rules (applets) to connect a variety of apps and devices together.

Generally, it helps users do more with their apps and devices.

The *ALEXA* device supports *IFTTT*, and it can trigger the *IFTTT* "rules" you have activated.

To illustrate:

If you ask *ALEXA* to find your phone, *IFTTT* can trigger your phone to ring. Or if you'd like have your *android* phone muted at bedtime, that can be done also.

New applets can be created, or you can choose from applets that already exist from other *IFTTT* users.

1.) If you haven't done so already, go to the *IFTTT* website and sign up.

2.) On the *IFTTT* website, find the *ALEXA* app by typing it into the search bar.

3.) Select the *Connect* tab when the *Amazon ALEXA* page comes up.

4.) Sign in to your *Amazon* account.

After signing in, your *Amazon* account should now be linked to your *IFTTT* account.

You have the option to remove the link between *ALEXA* and *IFTTT* at any time by visiting *Manage Login with Amazon*.

How to Create an Applet

1.) Click *invent your home* at the top-right corner of the screen on your *IFTTT* page.

2.) Select *New Applet*. Then click + *This*. *If This* is the trigger part of the process.

3.) On the *Choose a service* page, type *ALEXA* into the search bar and select it.

4.) You should now be on the *Choose trigger* page. If you'd like to customize the wording, scroll down to the box that reads, *Say a specific phrase*. For example, to turn off the lights, you might want to say, "Power off," instead of "Turn off the lights," so that's what you would type in.

5.) After choosing a phrase to trigger the action, select *Create trigger*.

6.) Click on + *That*. *If That* is the action part of the process. Then choose an action. For example, if you are using *WEMO* light bulbs and you'd like to power them on and off through *ALEXA*, search for *WEMO* in the search bar. In this case, you would then go to *WEMO lighting* and select *Connect*.

Although different apps have different connection methods, most of them are rather similar.

7.) After it's connected, go back and finish selecting an action. Following the previous example, if you have chosen *WEMO* light bulbs, you can now choose what you'd like to happen (Dim the light, Dim a group of lights, etc.) Then select the *Create action* tab.

8.) Select *Finish*.

To power the lights off, you would have to repeat the above steps. This time, on the *Create trigger* page, you can type in an "off" trigger, such as, "Power off."

There are plenty more things you can do, but the process generally remains the same. Scroll through the *IFTTT* lists and find something that interests you. When you are ready to set something up, follow the above steps outlined in this chapter.

How to Enable Voice-View Screen Reader

The *Voice-View Screen Reader* reads out loud the items you touch on the screen. It will also describe the actions you make on screen.

Enable *Voice-View Screen Reader*

1.) Press and hold the **Mic/Camera** button until you hear an alert.

2.) Hold two fingers (slightly apart) on the screen for approximately five seconds.

You can also:

1.) Go to **Settings**.

2.) Select **Accessibility**

3.) Go to **Voice-View Screen Reader**.

4.) Select **Voice-View**

<u>How to Enable and Use *Screen Magnifier*</u>

T he *Echo Spot Screen Magnifier* allows you to enlarge items on the screen.

1.) Go to **Settings**.

2.) Select **Accessibility**.

3.) Select **Screen Magnifier**.

Once enabled, *Screen Magnifier* can magnify the screen and adjust the zoom level.

Here's how it works:

Magnify the screen: Triple-tap the screen with one finger.

Temporarily magnify the screen: Triple-tap the screen with one finger and hold it on the screen. To pan, drag your finger around the screen.

Pan: Drag two fingers (slightly apart) across the screen.

Change zoom level: When the screen is magnified, pinch inward or outward with two fingers.

ALEXA Command and Request List

Basics

" ALEXA, turn up the volume. "

" ALEXA, turn down the volume. "

" ALEXA, let's chat. "

" ALEXA, stop. "

" ALEXA, go to sleep. "

" ALEXA, help. "

Music

" ALEXA, next song. "

" ALEXA, skip song. "

" ALEXA, previous song. "

" ALEXA, pause in [room name]. "

" ALEXA, resume in [room name]. "

" ALEXA, play the next track in [room name]. "

" ALEXA, louder in [room name]. "

" ALEXA, quieter in [room name]. "

" ALEXA, set the volume to [volume number or percentage] in [room name]."

"ALEXA, mute [room name]."

"ΛLEXA, turn it up in [room name]."

"ALEXA, what's playing in [room name]?"

"ALEXA, play music by [artist]."

"ALEXA, what's this song?"

"ALEXA, buy [album name] by [artist's name]."

"ALEXA, play the top songs this week."

"ALEXA, play my [playlist name] playlist."

"ALEXA, shuffle my new music."

"ALEXA, shop for new music by [artist's name]."

"ALEXA, play the [station name] on [music service name]."

"ALEXA, add this song."

"ALEXA, who sings the song [song title]?"

"ALEXA, who is in the band [band's name]?"

"ALEXA, sample songs by [artist]."

To-do lists

"ALEXA, add [item] to my shopping list."

"ALEXA, create a to-do list."

"ALEXA, put [task] on my to-do list."

"ALEXA, I need to [task]."

Shopping on *Amazon*

"ALEXA, add [item] to my cart."

"ALEXA, buy [product]."

"ALEXA, order [item]."

"ALEXA, reorder [item]."

"ALEXA, where's my stuff?"

"ALEXA, track my order."

Smart Home

"ALEXA, discover my *smart home* devices."

"ALEXA, BLUETOOTH."

"ALEXA, connect to my phone."

"ALEXA, is the front / back door locked?"

"ALEXA, lock the front / back door."

"ALEXA, turn on the lights."

"ALEXA, turn on the TV."

"ALEXA, raise the temperature [number] degrees."

"ALEXA, set the temperature to [number]."

"ALEXA, what's the temperature in here?"

"ALEXA, what's the thermostat set to?"

"ALEXA, make the living room [color]."

"ALEXA, turn the desk lamp to [color]."

"ALEXA, turn on the hallway light."

"ALEXA, turn on *Movie Time*."

"ALEXA, dim the living room to [percentage]."

"ALEXA, set the fan to [percentage]."

Weather

"ALEXA, what's the weather in [name of city]."

"ALEXA, what's the temperature?"

"ALEXA, what will the weather be like in [name of city] tomorrow?"

"ALEXA, what's the extended forecast for [name of city]."

"ALEXA, is it going to rain today?"

"ALEXA, will it snow tomorrow?"

"ALEXA, will I need an umbrella today?"

Traffic and Local Information

"ALEXA, how is traffic?"

"ALEXA, what's my commute like?"

"ALEXA, what are the business hours of [venue name]?"

"ALEXA, what [venues] are nearby?"

"ALEXA, what time is the movie, [film name] playing?"

"ALEXA, find the address for [place]."

"ALEXA, is [venue] open?"

"ALEXA, ask *UBER* to request a ride."

News

"ALEXA, what's in the news?"

"ALEXA, give me my flash briefing."

"ALEXA, open [publication name]."

"ALEXA, pause."

"ALEXA, next."

"ALEXA, previous."

Sports

"ALEXA, give me my sports update."

"ALEXA, what was the score of the [name of team] game?"

"ALEXA, did the [team's name] win?"

"ALEXA, when do the [team's name] play next?"

Alarm Clock

"ALEXA, set an alarm for [time]."

"ALEXA, when's my next alarm?"

"ALEXA, snooze."

"ALEXA, set a timer for [time length]."

"ALEXA, set a second timer for [time]."

"ALEXA, cancel my alarm for [time]."

"ALEXA, what time is it?"

"ALEXA, cancel all alarms."

"ALEXA, set a repeating alarm for [time] [days]."

Calendar

"ALEXA, what's the date?"

"ALEXA, add an event to my calendar."

"ALEXA, add a [time and event] to my calendar.

"ALEXA, what's on my calendar today?"

"ALEXA, what's my next appointment?"

Knowledge

"ALEXA, how tall is [name of mountain]?"

"ALEXA, how deep is [name of ocean]?"

"ALEXA, what's the capital of [place]?"

"ALEXA, what's the population of [place]?"

"ALEXA, who wrote [name of book]?"

"ALEXA, what's the definition of [word]?"

"ALEXA, how do you spell [word]?"

"ALEXA, what's [number] times [number]?"

"ALEXA, [number] factorial?"

AUDIOBOOKS

"ALEXA, play [book title] on *Audible*."

"ALEXA, pause."

"ALEXA, resume."

"ALEXA, next chapter."

"ALEXA, previous chapter."

"ALEXA, go to last chapter."

"ALEXA, go to [chapter number]."

Just for Fun

"ALEXA, tell me a joke."

"ALEXA, sing a song."

"ALEXA, tell me a story."

"ALEXA, play a game."

Troubleshooting

Many problems can be solved by simply restarting (unplugging the device and plugging it back in) the *Echo Spot*.

But when restarting *Echo Spot* does not correct the issue, there are also some other things that can be tried.

ALEXA App Doesn't Seem to Work

• Confirm your device meets the requirements

• Restart your phone

• If you are using a web browser, close the web browser, then reopen it

• Close the app, then reopen it

• Uninstall the app, then reinstall it

Problems with ALEXA Skills

• Disable the skill, then re-enable it

Streaming Issues

• Verify your internet connection is at least 512 KBPS (0.51 MBPS)

• Turn off other devices that might be absorbing the bandwidth

• Move the *ALEXA* device closer to the router and modem

• Move the *ALEXA* device away from microwaves and other possible sources of interference

• Restart the router and modem

ALEXA Isn't Understanding The Words You're Saying

• If there is background noise, wait for it to clear before speaking to *ALEXA*

• If the *ALEXA* device is on the ground, try moving it to a higher location

• Be more specific

• Use voice training. See the *How to Teach ALEXA Your Voice* chapter.

BLUETOOTH Issues

• Verify your *BLUETOOTH* device uses a supported profile (A2DP SNK, A2DP SRC, AVRCP)

• Move the *BLUETOOTH* device away from microwaves and other possible sources of interference

• Make sure the *BLUETOOTH* device is close enough to the *ALEXA* device when you pair it

• Check the batteries and replace them if necessary

• Clear all the *BLUETOOTH* devices, then restart *ALEXA* and the *BLUETOOTH* device

To clear the *BLUETOOTH* devices with *Echo:*

1.) Open the *ALEXA* app.

2.) Go to the menu and select *Settings*.

3.) Select your device.

4.) Select **BLUETOOTH**.

5.) Select a device from the list.

6.) Select **Forget**.

7.) Repeat the process for all other *BLUETOOTH* devices.

Don't forget to restart the *ALEXA* device and the *BLUETOOTH* device when you are finished clearing the devices.

Smart Home Camera Issues

• Verify the camera is compatible with *ALEXA*

• Verify the *smart home* camera is powered on

• Check the battery or power supply

• Verify you have completed the setup process through the camera manufacturer's website or companion app

• Check for available software updates and install them if necessary

• Check your camera's network settings

• Restart your devices

<u>How to Use The *ALEXA* Voice Remote (Sold Separately)</u>

T he *ALEXA* Voice Remote is a voice-enabled, battery-powered remote that features a directional track pad, a microphone, and a *talk* button. It allows you to rapidly control audio playback on the *Echo* device.

It must be purchased separately.

Before it is able to be used, it must be paired with the *Echo* device.

<u>Pairing *ALEXA* Voice Remote with *Echo*</u>

1.) Open the latch on the battery door of the remote. The latch can be opened by pulling down on the battery door, then pulling the door away from the remote.

2.) Insert two AAA batteries, then close the latch.

3.) Open the *ALEXA* app.

4.) Go to the menu, then select *Settings*.

5.) Press and hold the **Play/Pause** button on the remote for approximately five seconds, then release it.

The *Echo* device should now search for the remote and connect it within forty seconds or so.

ALEXA will say, "Your remote has been paired," when the remote is discovered by the device.

How to Reset *Echo Spot's* Screen

If *Echo Spot* becomes unresponsive and restarting the device doesn't help, the device can be reset.

Important Note: Do not reset or power off the device if you are setting it up for the first time.

Reset *Echo Spot*

1.) Say, "Go to settings," or use a downward motion to swipe the upper portion of the screen and select **Settings**.

2.) Select **Device Options**.

3.) Select **Reset to Factory Defaults**. This will erase all of your personal information and settings on the device.

After the device resets, the setup process must be completed again.

More from Emery H. Maxwell

Fire HD 10 Tablet Manual, available at all *Amazon* stores, including <u>US.</u>